OIL RIG

A Comprehensive Guide On How To Find A Job In Oil Rig

OIL RIG

A Comprehensive Guide On How To Find A Job In Oil Rig

TABLE OF CONTENTS

INTRODUCTION

Many of this world's largest and most profitable oil reserves lie beneath the seabed, from the Persian Gulf to the North Sea. Meanwhile, as drilling techniques and technologies improve, our ability to extract undersea oil grows.

As a result, oil and gas companies constantly look for talented and adventurous individuals to live and work on offshore oil platforms worldwide.

Do you want to join them? As you might expect, many oil-based jobs require strong STEM skills and higher education in a related discipline. However, an engineering background is not required: you can work on an oil platform as a project manager, an HSE professional, or a roustabout, for example.

Continue reading to learn more about the jobs available on an offshore oil rig, some of the current global hotspots, and the qualifications you'll need. We'll also advise you on where to look for your next oil platform job and whether to apply through an oil company or a recruitment agency.

What Is the Role of an Oil Rig Worker?

Your responsibilities as an oil rig worker include assisting with oil and gas drilling and extraction operations on an offshore oil platform. You may be assigned a task on the rig, or your responsibilities may be more general. Oil rig positions such as derrick man and driller work directly with the drilling and extraction equipment. You could also do catering, cleaning, or providing medical services.

WHAT TYPES OF JOBS CAN BE FOUND ON AN OFFSHORE PLATFORM?

It is not uncommon for an oil platform to have a crew of up to 200 people, all of whom must live and work in close quarters for weeks at a time.

This will include offshore installation managers, engineers, and laborers, as well as a variety of support personnel to ensure that life on the platform is safe and comfortable and that environmental risks are minimized. Examples of oil rig jobs include:

Engineers of structures

Oil platforms have intricate structural designs. They must be able to continue drilling and pipeline operations in potentially harsh and unforgiving conditions. As a result, structural engineers are essential and high-demand crew members.

Engineering for drilling

Drilling engineers, as one might expect, play an

essential role in subsea oil drilling operations. They assist in developing and implementing drilling programs, monitor efficiency and safety, and ensure that work is done according to environmental protection standards.

Engineering of pipelines

Pipeline design, monitoring, and maintenance is another vital engineering role on an oil platform. After the oil is extracted from the seafloor, skilled engineers work to ensure that it is transported and stored safely, efficiently, and risk-free. They are also in charge of pipeline maintenance and repairs.

Project administration

Project managers are in charge of large-scale construction and engineering projects, including their planning, implementation, and monitoring. They are in order of the scopes and schedules, ensuring that resources are available for the work to be completed by best practices, and ensuring that outcomes and performance targets are met. Project managers on oil platforms frequently have an engineering background and may also be certified in PMP or PRINCE2.

HSE

Oil platform workers face hazardous conditions daily, so health, safety, and environmental (HSE) specialists are also essential crew members. An HSE professional on an oil platform will manage risks associated with heavy and electrical equipment, structural stability, flammable materials, and other factors.

Roustabouts are in charge of general labor such as deck maintenance, basic repair jobs, and offloading boat supplies. A roustabout on an oil platform does not require a degree, making it a possible entry-level position for someone interested in the industry. To work offshore, however, you must be trained and certified.

WHERE CAN I WORK ON AN OFFSHORE PLATFORM AROUND THE WORLD?

As previously stated, some of the world's largest oil fields are located offshore. Here are some examples:

• The North Sea - The North Sea contains Europe's most significant oil and natural gas reserves. Active producers include the United Kingdom, Norway, France, Denmark, Germany, and the Netherlands.

• The Gulf of Mexico - Offshore drilling off the coasts of Texas, Louisiana, Mississippi, and Alabama accounts for a sizable portion of US crude oil and natural gas production.

• The Gulf of Iran - The Persian Gulf is well-known for offshore oil, containing the world's largest oil field. It contributes significantly to the great economies of the United Arab Emirates, Saudi Arabia, and Qatar, among others.

Are you interested in working abroad but unsure what to expect?

Most oil and gas companies (as well as their recruitment partners) rely heavily on ex-pat labor. If you find a job that requires you to relocate across the globe, we can assist you with the relocation process.

WHAT QUALIFICATION WILL I REQUIRE TO WORK IN OFFSHORE OIL AND GAS?

Not all oil platform jobs necessitate a bachelor's degree or higher. To work as an offshore engineer, you must have a bachelor's degree or equivalent in mechanical, structural, or civil engineering. Many more technical positions in project management and health and safety on oil platforms may also require comparable credentials.

In addition, completing a course in offshore work is a legal requirement in many nations. For instance, the Basic Offshore Safety Induction and Emergency Training Certificate (BOSIET) is required to work offshore in the UK. While looking for work, many candidates do this on their own time and at their own expense, though some companies offer sponsorship.

Aside from that, all offshore workers will need to demonstrate skills such as:

- Excellent communication abilities (and the ability to communicate with workers from different disciplines and backgrounds)

- Capability to work as a member of a close-knit team (often living and working together nearby for weeks at a time)

- The ability to work under duress (especially in conditions that may be stressful and challenging, such as extreme weather and long working hours)

- Extensive attention to detail (especially when it comes to their health and safety and the safety of others)

- You should be 18 years old and have completed emergency response training.

- Apprenticeships are available but not required for unskilled workers.

- Engineering qualifications are frequently required for leadership positions.

- Entry-level training for engineering graduates

HOW CAN CANDIDATES PREPARE FOR OFFSHORE OIL RIG JOBS?

While it is true that oil rig jobs provide rewarding career opportunities, the work is mainly labor-intensive and may require working for extended periods. Before applying for offshore jobs in Texas, aspiring candidates should prepare for the job interview and be aware of the eligibility requirements. Candidates must have a high school degree to be considered for entry-level positions. This article provides several valuable tips to help candidates prepare for offshore oil rig job interviews.

Learn More About The Industry

Candidates should stay updated on the latest news and events in the oil and gas industry by regularly visiting websites that cover these. This will assist them in learning more about the career opportunities in this field and determining which potential locations around the world are likely to see growth.

Improve Your Physical Fitness

Exercise and physical fitness can help candidates increase their chances of landing an oil rig job. They can strengthen their muscles and increase their weight-lifting ability by working out in a gym or following a rigorous workout plan. A roustabout is expected to handle rig maintenance, machinery repair, and other physically demanding work on an oil rig. "Offshore roustabouts handle cargo from supply ships and helicopters, keeping the rig platform clean and repairing and maintaining machinery and equipment." To be considered for a roustabout position on an offshore oil rig, you must be able to lift and carry almost 100 pounds."

Obtain the Required Credentials

Candidates seeking employment on offshore oil rigs in the Gulf of Mexico must obtain a transportation worker identification credential from the United States Department of Homeland Security. This will help them qualify for offshore jobs with leading Gulf of Mexico companies.

In addition, candidates must be willing to undergo a

physical examination, a drug and alcohol test, and a background check. Only after passing these tests will candidates be able to apply for rewarding jobs on oil rigs.

Candidates seeking entry-level positions on oil rigs should be aware of some strategies for better preparing for job interviews and pursuing rewarding career opportunities.

HOW TO WORK ON AN OIL RIG

A manned platform used for land or offshore drilling is known as an oil rig. An oil rig worker is a person who works on either platform and is in charge of various tasks related to safely drilling for oil. Most workers' schedules consist of 14 to 21 days of work before being given time off. In addition to a wage, most offshore oil rig workers are provided with food and lodging and some travel expenses. Because oil rig managers value experience, getting your first job on an oil rig may be the most challenging part of the process. Discover how to work on an oil rig.

1. Make sure you meet the minimum requirements for employment on almost any oil rig. They are as follows:

- You are over the age of 18.

- You are in excellent physical condition. Mental fitness is also essential. Before you can be hired, you must pass a physical exam.

- You are a nonsmoker who can refrain from drinking alcohol during your long 14 to 21-day shifts.

- You are willing and able to work the odd hours of an oil rig worker. You should be ready to work nights and be able to work long days without a break.

2. Work as a mechanic to gain experience. Mechanical jobs are typical among oil rig workers. You can also pursue training as an electrician, cook, medic, or engineer to become an essential auxiliary oil rig crew member.

3. Make contact with friends, family members, or acquaintances who work on oil rigs. Ask around if you don't know anyone. This contact will assist you in understanding the work and will provide you with valuable industry contacts.

Look at industry chat boards online if you can't find any connections to current oil rig workers. You may find helpful information on the best and worst employers and who is currently looking for workers. Remember not to take all Internet chat seriously because many people use it to complain or argue.

4. Investigate the oil industry. Take a community college or online course, or simply go to the library and check out books about oil drilling, oil workers, and the most up-to-date information on petroleum industry changes and regulations.

You should be able to demonstrate to employers that you are interested in the petroleum industry and want to put in the effort required to be a valuable employee. If you are well-versed in essential topics, you will not appear to be such a novice. You can also question your oil rig contacts about topics or employers to research.

5. Look for job postings in newspapers and online. To find current listings for oil rig workers, go online to large job search sites such as Monster, CareerBuilder, and Indeed. Examine those listings to ensure you have the qualifications to begin working on an oil rig.

Oil rig jobs may have the following titles: Derrickman, Safety man, Driller, Assistant Driller, Sub Sea Engineer, Storekeeper, Crane Operator, Mechanic/Electrician, Rig Welder, Barge Engineer, Rig Medic, Toolpusher, or Mudman. People working hard labor jobs on oil rigs are sometimes called "roustabouts."

6. Look for job postings in newspapers and through oil industry contacts. Look for entry-level positions that include on-the-job training.

7. Make your resume and cover letter. List all the jobs you've held that required skills that could be transferred to an oil rig. Highlight any jobs in which you worked as a laborer, mechanic, or electrician.

A cover letter is typically a half-page to a full page of text that describes in prose why your experience qualifies you for the job.

A resume should be about one page long, but it can be up to two pages long. Concentrate on your qualifications and any accolades you've received for being a quick learner, team player, or dependable employee. Target your professional resume to the job you're applying for, and tweak it slightly for each one. Look for work. Apply for several jobs simultaneously because entry-level jobs will have more competition.

9. Seek Basic Offshore Safety Instruction and Emergency Training (BOSIET) certification. This is

required for work in the United Kingdom, the Netherlands, or Denmark. [4] You should also think about getting a Helicopter Underwater Escape Training certificate.

These training courses can be found at marine safety academies or schools. A marine safety school is available at many seaside colleges in the United Kingdom and the United States.

10. Apply for your Transportation Worker Identification Credential (TWIC) if you intend to work in offshore drilling in the United States. To obtain an application, go to twicprogram.tsa.dhs.gov. To apply, you must go to a TWIC enrollment facility.

Fingerprints, biographic personal information, identification documents, and photographs will be required. The Transportation Security Administration (TSA) will review the information and notify the enrollment center of the TWIC approval.

WHAT POSSIBILITIES DOES LINKEDIN GIVE?

LinkedIn provides a unique opportunity to directly reach company decision-makers. All you have to do is define who you are looking for, use the appropriate search criteria, and contact the individuals. Contact them and inquire about the required qualifications and experience and any additional training needed to apply for employment. It is worthwhile to consider building relationships on LinkedIn to achieve the objectives we set for ourselves.

Every company and agency that employs oil rigs has its own Linkedin company profile. You can learn about the company's history and current events by logging in to the website and typing its name into the search engine. You can search for job offers published by the company in the job tab. Using the search engine's advanced functions is worthwhile, such as entering keywords, job titles, specific companies, etc. After clicking "Apply," you will be taken directly to the company's website, where you can apply for a particular position. One of the options this tab provides is the ability to set up an alert for new job postings. When you have something useful, you are up to date.

Linkedin allows you to connect with people already working on oil platforms. Nothing prevents you from contacting them and directly inquiring about the possibility of work. Someone will provide you with helpful information and suggestions. Work on your contact network, look for people on LinkedIn with whom you have professional relationships and use the website to build a friend network.

Remember that your professional profile is your calling card!

IN A DRILLING ROBOT JOB

Oil has always been shrouded in secrecy for some reason. Many people see it as a place where hard work and many greases are required. However, working on a drilling rig differs from what you expect. Today, working on a drilling rig is like working in a high-tech office that is constantly evolving.

Maersk Drilling is a global provider of drilling services. Many of the rigs are based out of the North Sea region. As technology advances, so does the drilling rig, making it a challenging place to work. Since Maersk Drilling has made significant investments in automating its rigs, our employees must take courses and build competencies to learn how to use specialized equipment and new technology while not on the clock. This means that the function for which they are employed gives preference to candidates with a technical background, such as electricians and fitters. English is the primary language used in the industry, so fluency in the language is essential. Typically, new offshore workers begin their careers as junior employees. Before beginning work on the rig, all applicants must pass a series of safety training courses. It is the beginning of a long line of training sessions that will keep offshore personnel's skills and capabilities constantly improving.

MAERSK DRILLING: A PATH TO SUCCESS

If you have the right combination of abilities, competencies, and a positive outlook, we can help you achieve your career goals.

A WORKPLACE IS OPEN 24 HOURS A DAY. To keep the drilling rig running 24 hours a day, seven days a week, the crew must work two 12-hour shifts. There is a two- to four-week "hitch" period of work on the rig, followed by a few days of downtime.

TRAINING

All Maersk Drilling crew members must attend regular training sessions on technology and safety. For example, all employees working in the Drilling Section must pass a good control course, which includes a test on the advanced drilling simulator at Maersk Training in

Svendborg, at the very least, from the position of Assistant Driller and upwards. To ensure that proper procedures are followed in the event of a crisis, training is essential.

THREE PARTS OF THE RIGS

The rig's sections are divided into four main categories: marine, drilling, maintenance, and administration. These categories are discussed in more detail on the following pages. The Offshore Installation Manager oversees the rig's four sections (OIM).

Offshore Installation Manager (OIM): In charge of the whole thing

It is the responsibility of the Offshore Installation Manager (OIM) to ensure that the rig and its workers are safe and secure. It's his responsibility to keep the client's agent well-informed about everything on the rig. In most cases, Maersk Drilling uses OIMs who have received their training on-site at the company. The OIM must have a Certificate of Competency as a Master or Chief Engineer and extensive offshore drilling rig experience. In the same way, a ship's captain is required to complete a long series of safety, environmental, and leadership courses; he is needed to do the same on a drilling rig.

SECTION OF THE MARINE

During Dynamic Positioning, the Marine Section is responsible for all aspects of safety on the rig, including the jacking system, crane operations, equipment stowing, and heli-landing and heli-take-off. The Barge Engineer is in charge of the Marine section, which includes the positions listed below.

ROUSTABOUT

Roustabouts, like Able-bodied Seamen or Odd-job Men on a ship, are the most common type of new hire in the Drilling and Marine Sections. Maintenance, cleaning, rust removal, painting, and steering the massive cranes on board are all tasks that the Roustabout assists with. He is ready to go up on the drill floor or to train as a Crane Operator when the Roustabout has mastered the drilling rig and its daily routines. He begins as a calming presence during meal and coffee breaks, as he does with everyone else.

The Operator Of A Crane

Each rig's crane operator is in charge of operating every piece of equipment on board. While loading and unloading supplies and equipment from supply boats and onto the ship, he serves as a foreman for the Roustabouts, who he supervises. The Crane Operator must be certified as an Offshore Crane Operator and have previous experience as a Roustabout to be considered qualified.

THE ENGINEER IN THE BARGE

Managing the drilling rig from a barge is the job of the Barge Engineer. He's in charge of everything that goes on the deck, including the upkeep of the ship's machinery. He is also responsible for the ship's safety, which includes conducting regular drills and holding weekly safety meetings. His role is comparable to that of a ship's captain, and he is expected to have held this position for at least a few years. A training period as an Assistant Barge Engineer is required before a Barge Engineer can be hired. It is necessary to have a Master's Certificate of Competency.

Automated Positioning Device

He reports to the SDPO, relieving him of his duties by operating the DPO's station. Maintaining and controlling the ship's load condition is the primary responsibility of the DPOs. The DPO constantly updates the weights and weight distribution of the rig to maintain its structural integrity. The power control system's second main task is to control and distribute the power.

Positioning Officer At The Highest Level

To keep the unit above the desired location in a safe and operational state, the Senior Dynamic Positioning Operator (SDPO) serves as the primary watchkeeper and system administrator. That isn't all: The SDPO also serves as the DPO's watchkeeper in the central control room, which is operated at all times by the two officers. Data entry errors are avoided thanks to the SDPO's meticulous attention to detail and production of accurate models and weather conditions. The SDPO serves as the unit's first mate while traveling. The Offshore Installation Manager is ultimately responsible for the SDPO (OIM).

AREA DUE TO DRILLING

The Drilling section of the rig is where the drilling-related work is done. Roughneck is the entry-level position, and the Senior Toolpusher is in charge of the team.

ROUGHNECK

The Roughneck, also known as the Floorhand, is tasked with performing the most physically demanding duties on a drilling rig, working on the drill floor. During his 12-hour shift, he must always be on his feet, with only brief breaks for meals or coffee. The blow-out preventer (BOP) – a massive safety valve mounted above the well – must also be monitored as part of the job. Equipment that will be used in a borehole must also be ready for use by Roughnecks. Typically, a Roughneck will work as a relief for a Derrickman for some time before advancing to the next position on the career ladder.

DERRICKMAN

During drill string assembly and disassembly, the Derrickman is in charge of handling the derrick's pipes. It is also his responsibility to ensure that the mud pumps work as efficiently as possible and that there is enough mud to lubricate and maintain the pressure in the hole when drilling begins. The ground must meet the exacting standards set forth by the mud engineer.

ASSISTANT DRILLER

The position of Assistant Driller can be obtained after working as a Derrickman for some time. A Driller's assistant is their right-hand man. On the drill floor, he's responsible for ensuring that all of the drilling equipment is in working order and ready to go when needed. He also serves as a conduit between the drill deck and the ship's bridge. The Assistant Driller is also responsible for training new drill floor employees. Driller is relieved of his duties during meal and coffee breaks.

DRILLER

An individual known as a Driller is responsible for actually drilling the wells. He has his own office and can control the entire drilling process. All the parameters necessary for the operation can be measured and adjusted here. Keeps an eye on how many pipes are in the hole and updates the progress report. To land a job as a Driller, one must have strong math skills. Every day, the Driller takes charge of the drill floor and words directly to either the Tourpusher or the Senior Toolpusher on duty for guidance and direction. If you work as a Driller for a long enough time, you may be promoted to Tourpusher.

Tour promoter/Lead Driller

As the well is drilled, the Lead Driller ensures that all safety regulations are followed. STP (Senior Toolpusher) is responsible for taking over the Lead Driller if necessary. The Lead Driller communicates with the company representative to ensure that progress and plans are understood. The lead driller can be promoted to STP after demonstrating their abilities.

Engineer in Training

As a mechanic or engineer, the Assistant Subsea Engineer (ASSE) is qualified for this position. The Subsea Engineer supervises and instructs the ASSE in its work. The blow-out-preventer (BOP) and the systems that control it are of primary concern. During the BOP, ASSEs should expect a tense and intense working environment with little room for error. Subsea Engineer promotion is the ultimate goal after a series of successes.

Engineer Specialized in Undersea Work

When the BOP stack is on deck, the Subsea Engineer (SSE) is responsible for preparing the work scope and plans. It will be the responsibility of the SSE to maintain and repair the BOP's control systems, which will be coordinated with the Subsea Supervisor in advance (SSS). Accordingly, it is expected that the SSE can identify and assign tasks based on the experience of its members. The Subsea Supervisor is responsible for overseeing the Subsea Supervisor's work. Subsea Supervisors will only be promoted to this position if they have a proven track record of leadership and technical proficiency.

Supervisor Of Subsea Personnel

The Senior Toolpusher is responsible for the Subsea Supervisor (STP). He is responsible for ensuring that the BOP stack and its control systems are working correctly.

He is responsible for ensuring that the BOP is tested following the intervals agreed upon by all stakeholders. He is in charge of maintaining and repairing the BOP and its control systems. A BOP crew member supervises the work scope when it is on deck. He's in charge of ensuring that everything needed for the job is procured, from spare parts to tools.

TOOLPUSHER, SENIOR

As a Senior Toolpusher, an employee is expected to be knowledgeable about drilling operations. Most employees with a decade or more experience with Maersk Drilling are the majority. Safety, environmental, and leadership courses are required for a Senior Toolpusher. The OIM's deputy is the Senior Toolpusher.

MAINTENANCE DEPARTMENT.

The Maintenance section looks after the machinery and drilling equipment on board. The Maintenance Supervisor is in charge of this section and is supported by several other employees, all of whom are listed below.

MOTORMAN

Every day, the Motorman lends a helping hand to the Mechanic. A Motorman must be a Certified Fitter or Mechanic or have completed coursework in a related field to be considered for the position.

MECHANIC

The Mechanic is responsible for maintaining and inspecting the rig's mechanical systems and equipment on board. Prior experience as a Motorman is required to land the job of Mechanic.

ELECTRICIAN

The Electrician is in charge of inspecting and maintaining all of the ship's electrical systems and equipment. The Electrician also performs maintenance and repair work on existing systems and new installations. He needs to be an Industrial Electrician or a Certified Electrician.

WELDER

According to the Maintenance Supervisor's instructions, the Welder performs repairs and modifications to the vehicle. He is also responsible for the repair and upkeep of all of the welding equipment. This position necessitates the possession of a Certified Welder diploma.

Software as a Service

The Maintenance Supervisor is responsible for the SAP planner's work. Maintaining data in SAP's P3M maintenance system is his primary responsibility. He flags urgent tasks and coordinates with the Logistics Coordinator regarding spare parts, planning, coordination, and support for all departments with data entry and extraction from the maintenance system.

Technician in Hydraulics

Maintaining drilling equipment, including tubular and pipe handling machines, falls to the Hydraulic Mechanic's remit. It is up to him to ensure that the work he has started is completed safely and in a high-quality manner. Hydraulic mechanics must have several years of experience working with hydraulic machines under their belts before entering the workforce.

The Engine Room Is In The Dark About This

The Maintenance Supervisor is responsible for the

Engine Room Responsible (ERR). Aside from ensuring everything needed to generate electricity is maintained correctly, he also provides that everything in his division is done in a safe and responsible manner. His primary responsibility is ensuring all applicable rules and regulations are adhered to in the engine room. As the head of his department, he assigns tasks to his team and ensures they have enough spare parts on hand in an emergency. Supervisor of Electronics Technicians Maintaining the proper functioning of all electronic and electrical equipment is the responsibility of the Electronic Technician Supervisor (ET Supervisor), who reports to the Maintenance Supervisor. When it comes to working, he ensures it is done safely and responsibly. He oversees the completion of tasks by his team members and makes confident that spare parts are readily available.

Technician In Electronics (Dynamic Positioning & Subsea)

The Electronic Technicians with specific training (ET DP & ET SS) are in charge of their respective systems. ET supervisors are briefed on system status by these workers. They must follow all applicable rules and regulations when performing system maintenance and repair. If possible, breakdowns should be avoided at all costs by a high level of care.

INSPECTION AND REPAIR CHEF

The drilling rig's technical manager is pictured here. Among his responsibilities are the upkeep and repair of the drilling equipment and machinery and any necessary technical modifications. To do his job effectively, the Maintenance Supervisor must be proficient in computers, as he is in charge of the drilling platform's electronic maintenance system. In addition, he must make sure that the rig is well-stocked with all necessary spare parts. He's in a similar position to a ship's Chief Engineer. Before being promoted to Maintenance Supervisor, a period of training as an Assistant Maintenance Supervisor is required. The Maintenance Supervisor position necessitates the possession of an STCW Certificate of Competency.

The Section Devoted To Management

The OIM is responsible for a large number of

administrative positions.

The Manager Of Logistics

The OIM is the one who oversees the Logistics Coordinator's work. The Logistics Coordinator is in charge of ensuring that the warehouse's computerized inventory control system runs smoothly. In addition to inspecting incoming materials, he follows up on requisitions from section heads.

MEDIC

There must always be a medical professional on board any drilling rig. Crew members will receive medical attention from this position, which is also responsible for maintaining a sufficient supply of medical and nursing supplies. The Medic is in charge of coordinating with the OIM if a patient needs to be evacuated for medical reasons. The Medic also performs personnel-related tasks, such as conducting safety briefings for new hires. To become a Medic, one must have a degree in nursing from an accredited college or university.

Officer In Charge Of Safety

The OIM is in charge of the Safety Officer's work. He assists the platform's management in ensuring that the unit's health, safety, and environmental (HSE) activities comply with applicable regulations. He is in charge of campaigns and initiatives aimed at improving workplace safety.

SECRETARY OF CONTROL

The Rig Administrator is responsible for all external communications. Additionally, he provides a daily weather forecast while keeping an eye on radio traffic, including that on the emergency frequency. In addition to a Certificate of Competency, candidates must have at least two years of seagoing or offshore work experience.

SHORT STORIES OF OIL RIG EMPLOYEES

Yasmin Abdel-Magied

When people ask about working on an oil rig, I lean in and tell them all my "war stories." But, in reality, it's just like real life. I graduated and began working as an oil rig worker with no experience. It was difficult for me to manage as a new girl with no experience on an oil rig with 150 highly experienced men.

"It's like another world working on the rig." On land, I am usually the only female among a group of 30 or so. You might have one other girl on-site at times, but rarely more than one. It's party time offshore! Are you talking about four women out of 150? It's insane. It's fantastic. Some of the jokes are hilarious."

Craig George

Initially, I worked offshore construction as a logistics/personnel clerk on oil platforms, so my job on an oil platform was technically considered extremely simple. In comparison to, say, a fitter or a welder's helper, I didn't have much "physical work" to do. My schedule was usually 2 hours longer than the shift pulled by the construction crew. So, if a shift was working 12-hour shifts, I'd be working 14. I was pulling about 20 hours if they were pulling 18 hours. I had to get up earlier in the morning to prepare the required paperwork for the daily work schedule; then, I had to close all the paperwork out, log everyone's hours, do the daily budget, and attend a "managers" meeting where the foremen/inspector/company rep/platform boss went over the day's activities and took notes, and various other tasks. It was difficult at first, but it is now well worth it.

Windham Mike

On an offshore platform or oil rig, life is generally good. I was exhausted when I first started my career. To begin, you must work 12 hours out of every 24. Half of the workers work, while the other half eats, relaxes, and sleeps. You will share a common room with one or three other employees. (There are two and four-man rooms.) The older rigs and platforms no longer have bunk beds; all beds are now at floor level. There is internet (albeit slow, but adequate for email), telephones, TV, and movies. The food is delicious if you have a good cook. The food is bad if the cook is bad. On offshore installations, bad cooks don't last long. Remember, this is not a place to relax or get away from the rest of the world. It is a work setting.

Brack David

When I was 18, I spent the whole summer working on a rig at mile 84 of the Alaskan Highway in northern British Columbia. The work was demanding, and I couldn't always do my fair share. Thank goodness they looked after me.

My most vivid memory is that EVERYTHING was difficult and weighed more than I did. The driller needed two hands to lift the 48″ wrenches.

And I was the only one of the 12 guys with all ten fingers, and I was on the verge of losing one. But because it was a camp job, I got room and board and 56 hours per week.

Cassanga Pearson

I was at university at the time. I met an engineer who

has spent many years working in the oilfield in various countries. Even though he was an Exploration and Production Engineer and I was a Drilling Engineer. I asked, "What are the advantages and disadvantages of being an engineer?" He stated the following to me:

- Being an engineer means that every day is a new challenge, and you must make quick decisions;

- You travel frequently; companies want to set their goals, and they sometimes don't care how much money it takes to achieve them. For example, one of my friends had to travel from Siberia to Houston to get a drilling bit, and the company spent nearly $10,000 (hotel, air ticket, pocket money, etc.);

- Good pay; pay varies depending on the position.

- The majority of drilling locations are in remote areas such as deserts, jungles, and the sea. Far from metropolitan areas, and if you work in the field rather than the office, you work on rotations, which can last two weeks, a month, or even longer.

Yes, he was correct. When you work in the oilfield, you can miss out on many special occasions, depending on your rotation. Graduation, New Year's, birthday parties, Christmas, and most importantly, you miss your family.

CONCLUSION

A single rig typically houses more than a hundred offshore workers. The interior appears to be a strange hybrid of a hotel and an office. There are shared cabins with shared bathrooms for staff accommodation. Some canteens serve meals, snacks, and cooking and cleaning services. Most modern rigs have recreational facilities such as gyms, cinemas, and game rooms.

Because of the unstable phone signal, it may appear that contacting the home is a difficult task. Except during duty hours, you can always access social media and calling/video apps like Skype and others.